discovermore

Exploring Physical Science

Matter

Marie Harts

Britannica Educational Publishing
IN ASSOCIATION WITH
THE ROSEN PUBLISHING GROUP

Published in 2024 by The Rosen Publishing Group, Inc.
2544 Clinton Street, Buffalo, NY 14224

Editor: Shannon Harts
Book Design: Rachel Rising

Photo Credits: Cover Sebastian Duda/Shutterstock.com; (series background) Dai Yim/Shutterstock.com; p. 4 liuhuaxuan/Shutterstock.com; p. 5 Solarisys/Shutterstock.com; p. 7 Shairaa/Shutterstock.com; p. 7 CiEll/Shutterstock.com; p. 8 H_Ko/Shutterstock.com; p. 9 FotoHelin/Shutterstock.com; p. 10 BearFotos/Shutterstock.com; p. 11 AppleZoomZoom/Shutterstock.com; p. 12 l i g h t p o e t/Shutterstock.com; p. 13 Ahanov Michael/Shutterstock.com; p. 14 chingyunsong/Shutterstock.com; p. 15 Irishasel/Shutterstock.com; p. 16 onimate/Shutterstock.com; p. 17 Festa/Shutterstock.com; p. 18 concept w/Shutterstock.com; p. 19 Pavlo Lys/Shutterstock.com; p. 20 Shebeko/Shutterstock.com; p. 21 TappyLabs/Shutterstock.com; p. 23 firatturgut/Shutterstock.com; p. 23 Danijela Maksimovic/Shutterstock.com; p. 24 Eugene Onischenko/Shutterstock.com; p. 25 Asphalt STANKOVICH/Shutterstock.com; p. 26 StockEU/Shutterstock.com; p. 27 https://commons.wikimedia.org/wiki/File:Albert_Einstein_photo_1920.jpg; p. 28 MarcelClemens/Shutterstock.com; p. 29 Vadim Sadovski/Shutterstock.com.

Cataloging-in-Publication Data

Names: Harts, Marie.
Title: Matter / Marie Harts.
Description: New York : Britannica Educational Publishing, in Association with Rosen Educational Services. 2024. | Series: Discover more: exploring physical science | Includes glossary and index.
Identifiers: ISBN 9781641900577 (library bound) | ISBN 9781641900560 (pbk) | ISBN 9781641900584 (ebook)
Subjects: LCSH: Matter--Juvenile literature.
Classification: LCC QC173.36 H37 2024 | DDC 530.4--dc23

Manufactured in the United States of America

CPSIA Compliance Information: Batch #CWBRIT24. For further information contact Rosen Publishing at 1-800-237-9932.

Contents

What Is Matter? 4
Super Solids 6
Learn About Liquids 8
Floating Freely with Gases 10
How Matter Changes Physically 12
How Matter Changes Chemically 14
Amazing Atoms 16
Understanding Elements 18
Arranging Elements 20
The Magic of Molecules 22
Fantastic Forces 24
Matter and Gravity 28
Glossary 30
For More Information 31
Index 32

What Is Matter?

Everything on Earth is made of matter—including you! Matter is found beyond Earth as well. Matter is anything in the universe that takes up space and has **inertia**. Air, water, rocks, and people are examples of matter. All matter is made up of tiny particles, or bits, called atoms.

One way of describing different types of matter is by their mass. The mass of an object is the amount of material that makes up the object. A bowling ball, for example, is heavy. It has more mass than a beach ball, which is filled with gas. Even though they are nearly the same size, the bowling ball is heavier because it has more mass than the beach ball.

While solids, liquids, and gases are the three clearly defined states of matter, there are some less clearly defined states of matter. Foam is one of them.

Objects with more mass have more inertia, meaning it is harder to change their state of motion. This is why it is harder to get a big, heavy ship moving than a smaller, lighter sailboat.

Matter comes in many different forms. There are three main forms, called states. These states are solid, liquid, and gas.

WORD WISE

Inertia is the property by which matter stays at rest or in the same straight-line motion unless acted on by an unbalanced force, or a force that is stronger in one direction. Put another way, objects with mass "keep doing what they are doing." This is called the law of inertia.

Super Solids

Have you ever left a cup of water outside on a freezing cold day? You may have noticed a while later that it became solid ice! Certain things are true of ice that are true of all solids. For example, ice is colder than its liquid form, water. If a piece of ice is placed in a cup, it keeps its own size and shape. Other solids, including rocks, books, and even individual grains of table salt, keep their size and shape too.

The atoms that form solids are held in place by an electromagnetic force. This is a force that pulls atoms close to their neighbors. The atoms always vibrate, or shake. Although they are moving, the atoms stay packed tightly together, in one position.

The force of attraction between the atoms of solids, such as those in a tree trunk, is strong. The force of attraction between the atoms of liquids, such as those in pond water, is more moderate.

compareandcontrast

Solids and liquids both have mass. Solids have a defined shape, whereas liquids take the shape of whatever container they are in, such as a can or bottle. What are some other differences and similarities between liquids and solids?

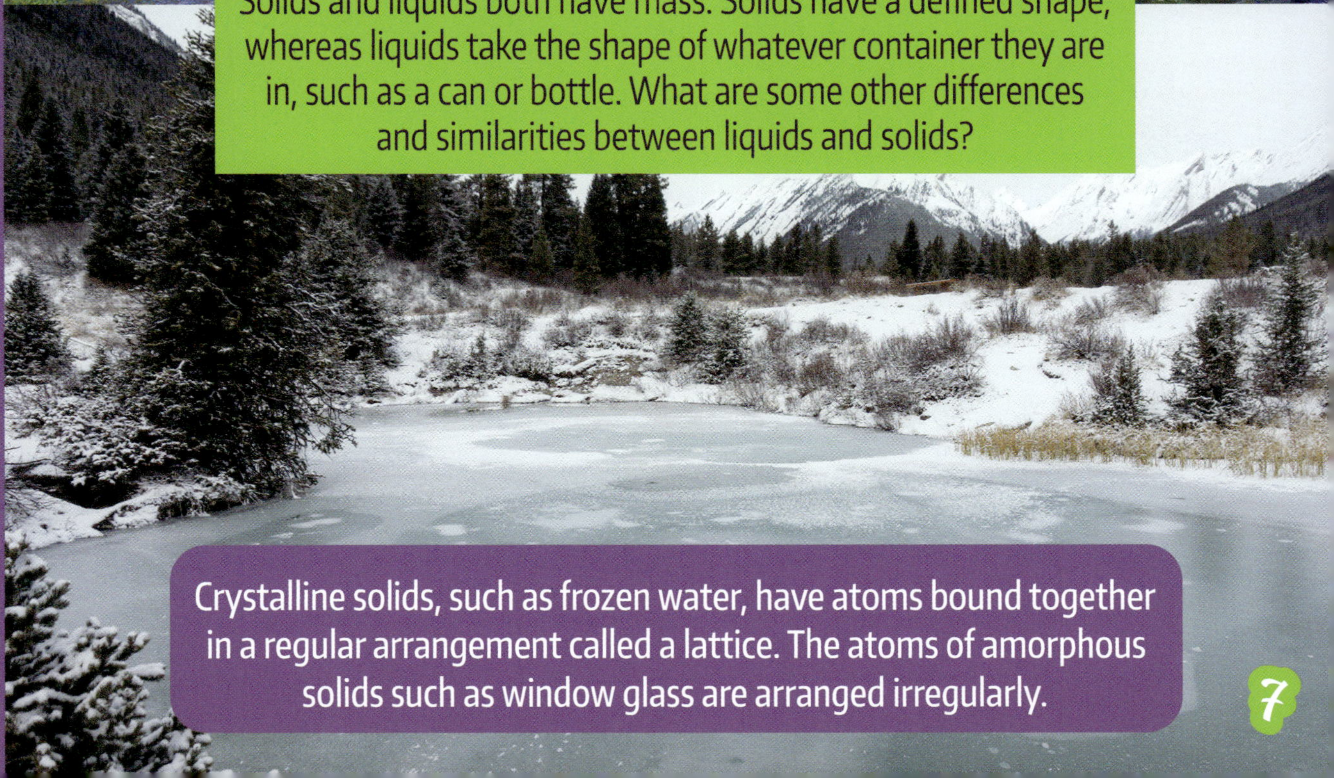

Crystalline solids, such as frozen water, have atoms bound together in a regular arrangement called a lattice. The atoms of amorphous solids such as window glass are arranged irregularly.

Learn About Liquids

Anything you can pour into a cup is a liquid—this may include your favorite drinks, such as milk, juice, and water. A liquid's shape depends on its container. Otherwise, it just runs all over. For example, milk changes shape when a person pours it from a carton into a glass. But the amount of milk stays the same. Liquid matter has a set volume regardless of the container. Volume is a way to measure the amount of space that matter takes up.

Mercury is the only metal that is a liquid at room temperature—that people know of!

Bubbles are round due to water's surface tension pulling water's bonded atoms into the tightest groupings possible. Of all shapes, a circle is the tightest possible grouping of particles because it has the least amount of outside area.

As with solids, electromagnetic forces also attract the atoms that make up liquids to each other. But the atoms are farther apart. There are fewer of them in the same space. This means the atoms have more room to freely slip around each other, but they are still attracted to each other.

Consider This

Surface tension is a liquid's ability to resist an external, or outside, force. This allows small objects to float on a liquid's surface. What are some ways people and animals (such as bugs) can use surface tension?

Floating Freely with Gases

Take a deep breath. You are breathing in matter in the form of gases, primarily nitrogen and oxygen. You can breathe in gases because they don't have either a set size or a set shape. The average distance between gas atoms is very large compared to their tiny size. So, the attracting forces between the atoms in a gas are weak. Many gases, such as nitrogen and oxygen, float freely in our atmosphere.

Have you ever wondered why balloons float? It's because they are filled with helium, a gas that's lighter than air.

Hot air balloons have burners that heat the air inside the balloon. This causes the air inside the balloon to become less dense than the surrounding air. This allows the balloon and basket to float!

Gas fills the entire volume of a container, unlike liquids or solids. If you put gas into a container, the atoms move freely until they hit the wall or another atom. You can't see gases, but like all other forms of matter, they still have mass!

Consider This

Can you think of ways to measure the mass of a gas? One way is to put the gas in a closed container, and then to measure the weight of the container on a scale. However, be careful to consider the container's weight and how the container is being weighted by the mass of the air surrounding it.

How Matter Changes Physically

Matter can be measured in ways that change depending on its form, or state. Matter can change its size, shape, or state without changing what it's made of! This is called a physical change. Properties such as how something feels, its color, and its smell are different for each kind of matter. Matter can change from one state to another at different **temperatures**.

You don't need special tools to measure some properties of matter, such as smell! Just put your nose close to an object and take a sniff!

A humidifier is a machine that can turn liquid water into water vapor. This adds moisture to the air that can help people with a stuffy nose break up mucus.

Heat makes liquid water evaporate, or turn into water vapor—a gas. Liquid turns into a gas at the temperature called its boiling point. The water vapor will become liquid again when it's cooled. Cooled down to 32°F (0°C), its freezing point, water will become solid ice. Ice will change back to liquid water if it's heated above its freezing point.

WORD WISE

Temperature is how hot or cold something is.

How Matter Changes Chemically

All matter is made of chemicals. A chemical is a substance that has a specific composition, or is made up of all the same "stuff." Examples of chemicals are water and chlorine. A chemical property tells how matter will change under special conditions. For example, iron rusts if it's left out in the rain. Paper and wood burn to ashes if they touch a flame. Burning and rusting are two examples of chemical reactions.

There are many examples of chemical changes in your everyday life. One is when you **digest** food.

Baking is an example of an irreversible chemical change. Once your cookies or muffins are made, you can't change them back into dough!

Chemical changes are important in many industrial processes. Oil, for example, can be changed to make plastic. Sand can be made into glass. When a material is changed but can still change back to its former state, it is said to have had a reversible change. An irreversible change means it is impossible for matter to go back to its former state.

WORD WISE

To digest is to break down food inside the body so that the body can use it. It involves the food matter being chemically changed into new, smaller substances.

Amazing Atoms

You've probably heard of or played with Lego blocks. Atoms are a bit like toy blocks in that they come together to create all types of matter! Each atom is made of smaller particles—electrons, protons, and neutrons. These are called subatomic particles. At the center of an atom is a nucleus. The nucleus has protons and neutrons stuck together. Protons carry a positive electrical charge, while neutrons have no electrical charge. Surrounding the nucleus is a cloud of negatively charged electrons.

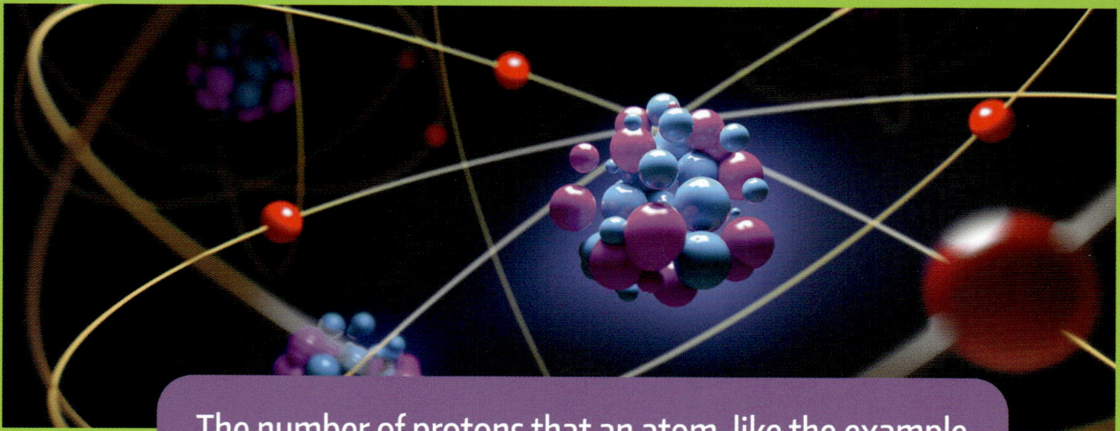

The number of protons that an atom, like the example shown above, has in its nucleus never changes.

It might be hard to imagine anything smaller than an atom, proton, or neutron, but quarks are even smaller! They are the particles that make up protons and neutrons.

Each type of atom has a different number of protons. This number of protons determines what kind of atom it is. For example, an atom with six protons is a carbon atom. Atoms that have the same number of protons, but a different number of neutrons, are called isotopes.

Consider This

Atoms with a neutral charge have the same number of positively charged protons as negatively charged electrons. What do you think happens to an atom's charge if it gains or loses a negatively charged electron?

Understanding Elements

Atoms are the smallest part of chemical elements. A chemical element is a basic substance. It cannot be broken down into simpler substances. Ninety-eight elements are found in nature. Scientists have created about 20 additional elements.

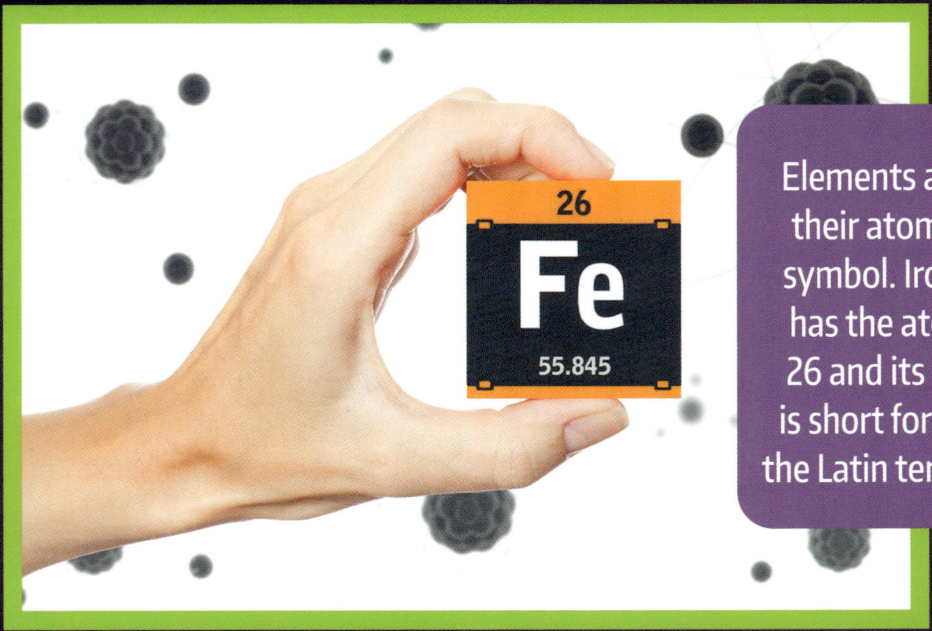

Elements are identified by their atomic number and symbol. Iron, for example, has the atomic number of 26 and its symbol is Fe. Fe is short for *ferrum*, which is the Latin term for "firmness."

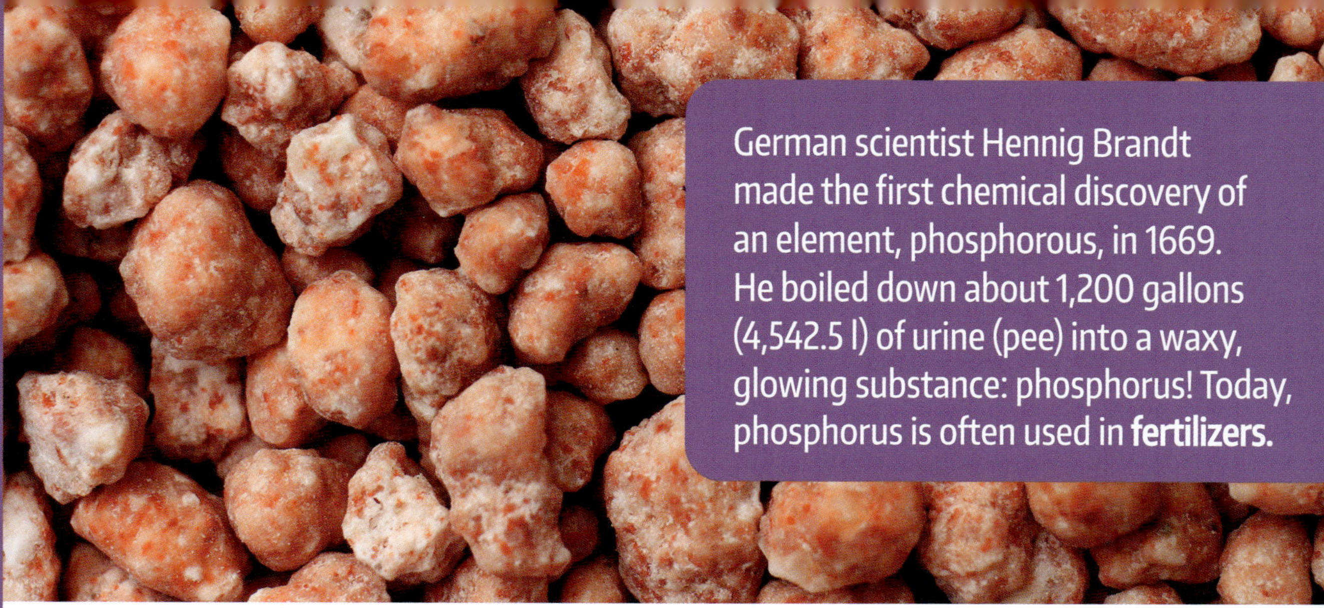

German scientist Hennig Brandt made the first chemical discovery of an element, phosphorous, in 1669. He boiled down about 1,200 gallons (4,542.5 l) of urine (pee) into a waxy, glowing substance: phosphorus! Today, phosphorus is often used in **fertilizers.**

Every element has its own symbol. The symbol for some elements is the first letter of their name. For example, H is the symbol for hydrogen and O is the symbol for oxygen. Other elements have a symbol with two letters. Examples include cobalt with the symbol Co and calcium with the symbol Ca.

Consider This

An element's atomic number is its number of protons. Its mass number is its number of protons plus its number of neutrons. Lithium has three protons and four neutrons. What is its atomic number and mass number?

WORD WISE

A fertilizer is something added to soil to help plants grow.

Arranging Elements

With so many elements in the world, it's important to have a system to arrange them. This system is called the periodic table. It arranges the elements in rows (left to right) and columns (top to bottom). In the rows, the elements are placed in order of their atomic number. The columns form groups of elements that have similar chemical properties. For example, certain gases are in one column and metals are in another. The periodic table helps **chemists** think about the elements and their properties.

Columns in the periodic table are called groups. Rows on the periodic table are called periods.

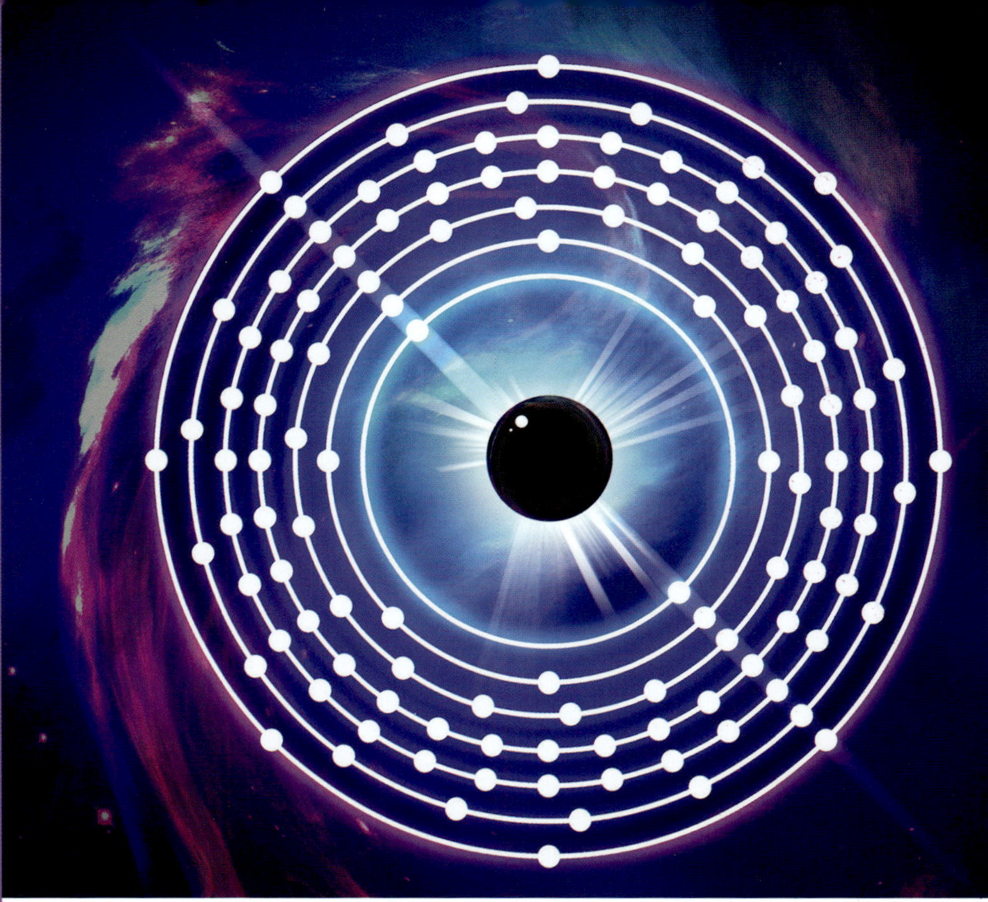

The known element with the highest atomic number is oganesson. It has an atomic number of 118.

Russian chemist Dmitry Mendeleyev developed the first periodic table in 1869. As scientists learned more about the elements, they have revised the table several times. The last update to the periodic table was in 2016, when four new elements were added.

WORD WISE

A chemist is trained in chemistry, which is the science involving the makeup, structure, and properties of substances, in addition to the ways substances change.

The Magic of Molecules

Molecules are made up of a group of two or more atoms. They're the smallest units that a pure substance can be divided into. For instance, a water molecule is the smallest unit that is still water. It's made up of two hydrogen atoms and one oxygen atom. But these atoms alone don't have the properties of water.

A strong chemical bond holds the atoms in a molecule together. Bonds form when atoms share electrons. Sometimes electrons are found either alone or in pairs in the outer part of atoms. When two atoms with unpaired electrons approach each other, the unpaired electrons may form a pair. Both atoms then share the pair. This creates a bond. Atoms tend to bond until they have eight electrons in their outer shell.

The sugar molecule called glucose is made up of six carbon atoms bonded into a chain. The molecule also includes 12 hydrogen atoms and six oxygen atoms.

Glucose - $C_6H_{12}O_6$

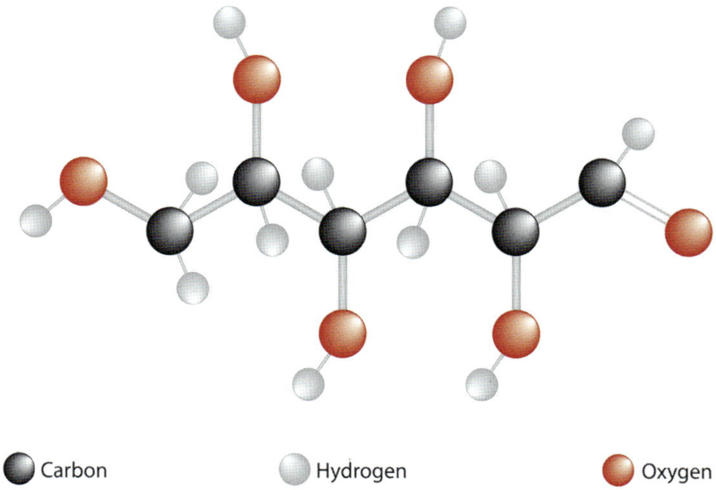

compareandcontrast

What are some similarities and differences you may notice between the glucose molecule and caffeine molecule? An example of a similarity is that both have hydrogen atoms.

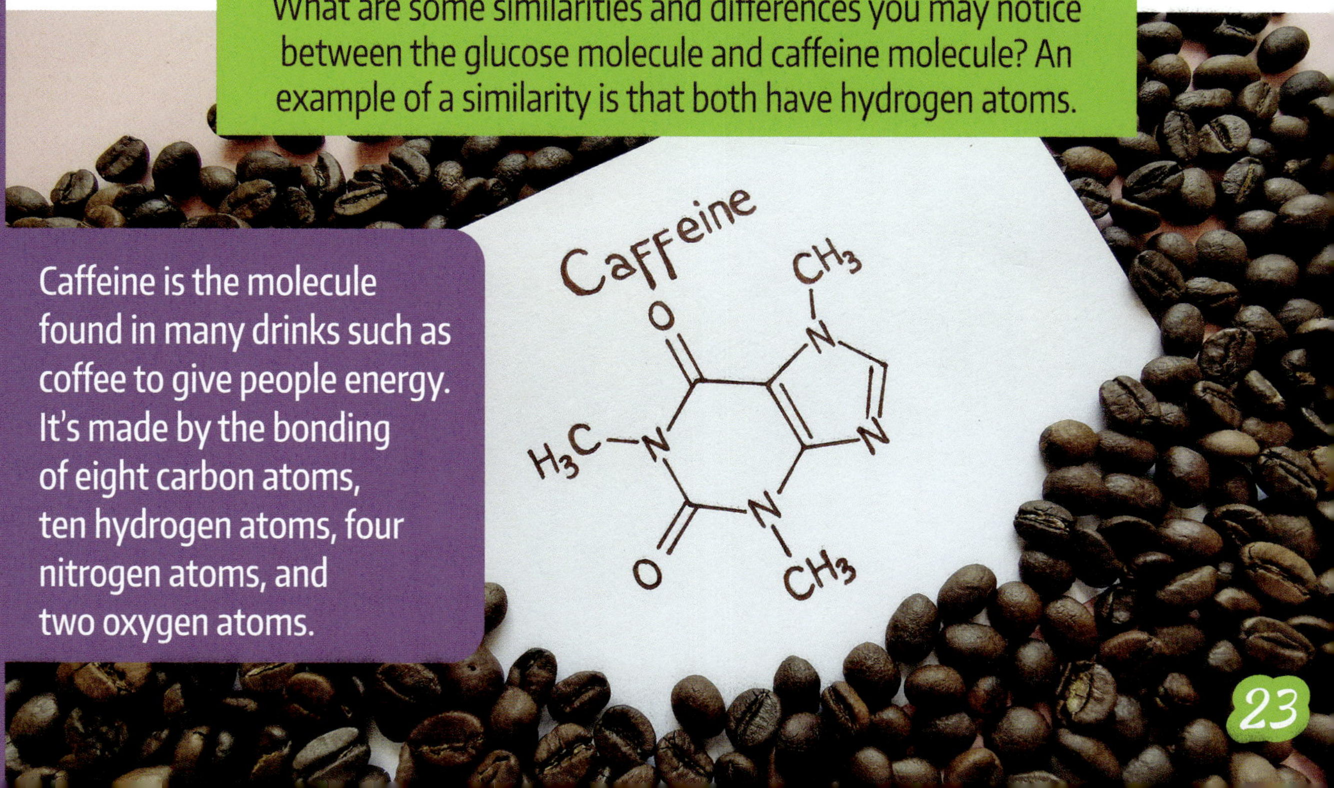

Caffeine is the molecule found in many drinks such as coffee to give people energy. It's made by the bonding of eight carbon atoms, ten hydrogen atoms, four nitrogen atoms, and two oxygen atoms.

Fantastic Forces

People who study matter and the forces that act on it are called physicists. In the late 1600s, English physicist Isaac Newton came up with three important laws about matter's motion:

1. The law of inertia (discussed on page five). An example of inertia that you may have experienced is when a car stops quickly, and you slide forward in your seat until you are stopped by your seatbelt. In this example, the unbalanced force that stops you from sliding forward is provided by your seat belt.

Newton's first law can be seen in hockey. The puck glides over the ice at the same speed until it hits an unbalanced force, either from another player's stick or the back of a net.

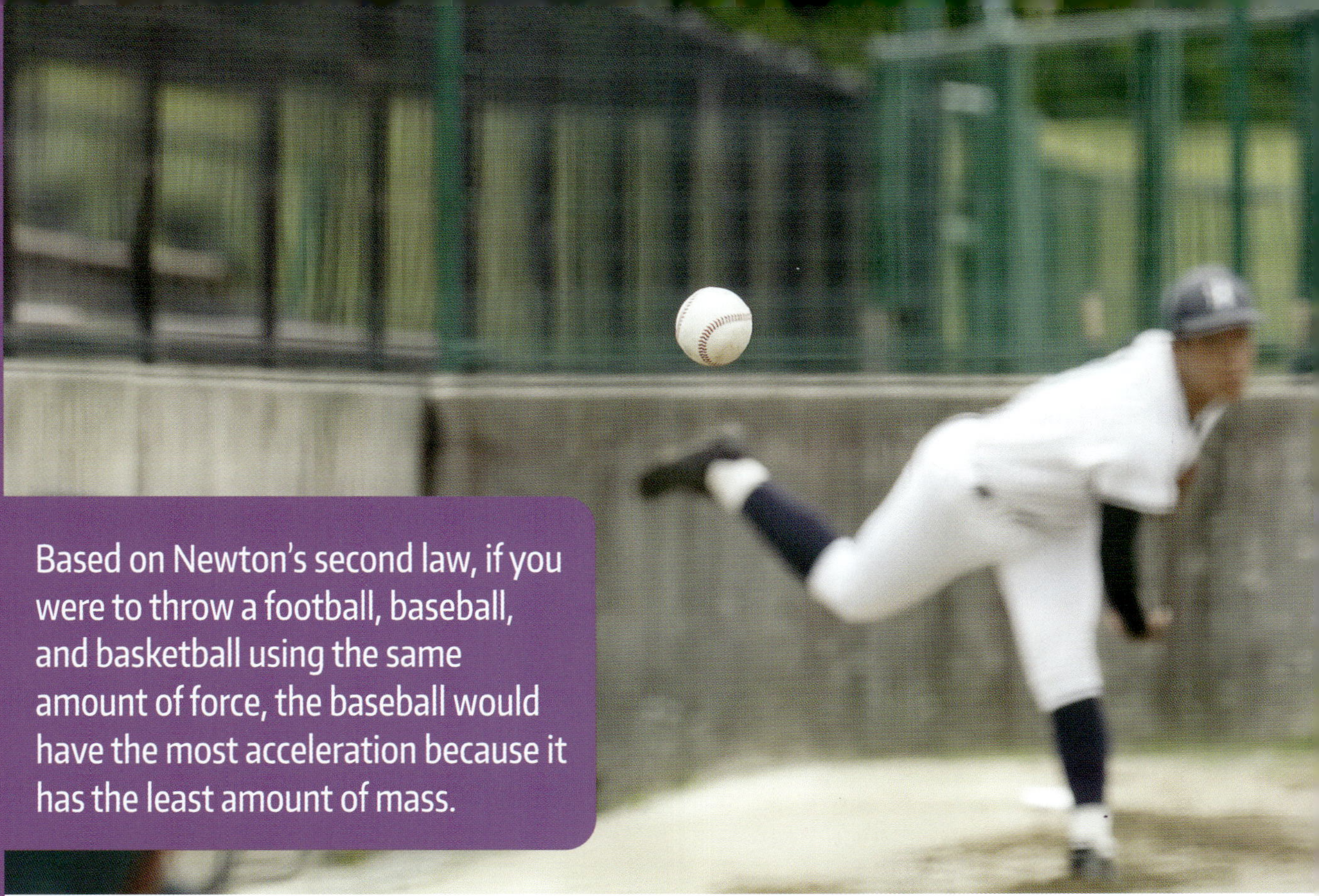

Based on Newton's second law, if you were to throw a football, baseball, and basketball using the same amount of force, the baseball would have the most acceleration because it has the least amount of mass.

2. An object's acceleration depends on the total force being applied to the object and the object's mass. If you would like to kick a soccer ball faster, you need to use more force when kicking the ball, based on this law.

Consider This

Newton's laws are important in all sports. What examples of the first and second law can you think of in sports that you have played or watched?

3. Newton's third law is called the law of action and reaction. It states that for every action involving matter, there's a reaction that uses an equal and opposite amount of force. When a person hits a ball with a tennis racket, the force of the racket causes the ball to move. At the same time, the force of the ball pushes backward on the racket, and the player feels the impact.

Cannons firing demonstrate Newton's third law. Each cannon exerts, or applies, a force on the cannon ball to shoot it out of the cannon, and the cannon ball exerts an opposite and equal force on the cannon.

Famous physicist Albert Einstein is considered one of the founders of quantum theory. Quantum theory is based on ideas about how energy and matter act at the atomic and subatomic levels.

While Newton's laws explain many of the ways we can see matter act, they don't apply to the **quantum realm**. This means that subatomic particles don't follow Newton's three laws of motion.

WORD WISE

The quantum realm is the way of seeing the universe as including very small particles and atoms. In the *Ant Man* Marvel comics and movies, the superhero Ant Man can enter a made-up quantum realm by wearing a special suit!

Matter and Gravity

All matter is pulled together by a force called gravity. Gravity is a pulling force that works across space. So, objects don't have to touch each other for gravity to affect them. For example, the sun pulls on Earth, even though it's millions of miles away.

Although Saturn is much larger than Earth, because it is made up mostly of gasses, it doesn't have much more mass than Earth. So, the surface gravity on Saturn is about 107 percent of Earth's surface gravity.

Astronauts seem weightless, but they have mass and are affected by Earth's gravity. They don't feel their weight because there is no ground beneath them to counteract the force of gravity—they are in a continuing state of freefall!

On Earth, gravity pulls objects toward the center of the planet. This makes objects fall and gives them weight. Weight measures the force of gravity between an object and the surface it stands on. When you measure your weight, you are measuring the amount of matter in your body, or your mass, and the force of gravity pulling you to Earth's center.

Consider This

The farther you are from Earth's center, the less you weigh due to the force of gravity. Do you think this means you would weigh less at the top of a tall mountain?

Glossary

acceleration The rate at which a moving object's speed changes over time.
amorphous Having no defined form.
atmosphere The mixture of gases that surround a planet.
attraction In physics, a force that pulls something toward something else.
chemical properties Characteristics of a substance that allow it to change into a different substance.
container An object used to hold something.
counteract To act against something; to lessen or prevent something's effect.
define To show or describe something clearly and thoroughly.
dough A mix of flour and water used to make baked goods.
electromagnetic force A term used to describe how both moving and still charged particles interact using magnetic and electric forces.
industrial Having to do with industry, or the process of making products using machines and factories.
lattice A frame or structure made of crisscrossed strips.
moderate Neither too much nor too little in size or amount.
moisture A small amount of liquid that makes something slightly wet or moist.
mucus A thick protective slime produced by the bodies of many animals.
nucleus The center of an atom; the part of an atom made up of protons and neutrons.
subatomic Either relating to the inside of an atom, or relating to particles that are smaller than an atom.
symbol A picture, shape, object, or letter that stands for something else.
unit An amount of length, weight, or volume that is used for counting or measuring.
universe The whole world, space, and things people experience.

For More Information

Books

Crane, Cody. *Matter*. New York, NY: Children's Press, 2019.

Gardner, Jane P. *Matter*. Minneapolis, MN: Bearport Publishing, 2023.

Peterson, Megan Cooley. *Matter*. North Mankato, MN: Pebble, 2020.

Websites

Britannica Kids: Matter
kids.britannica.com/kids/article/matter/353444
Find out more facts about matter here.

Kids in Motion: States of Matter
www.pbs.org/video/kids-motion-states-matter-gr-2-5/
Get moving and find out more about different states of matter with this video showing an interactive activity!

Science Trek: States of Matter
www.pbs.org/video/science-trek-states-matter/
Watch a fun video to learn more about states of matter here.

Index

A

action and reaction, law of, 26
atomic number, 18, 19, 20, 21

B

baking, 15
boiling point, 13
bubbles, 9

C

chemicals, 14

E

Einstein, Albert, 27
electrons, 16, 17, 22

F

force of attraction, 7

I

inertia, 4, 5, 24

M

mass, 4, 5, 7, 11, 19, 25, 28, 29

N

neutrons, 16, 17, 19
Newton, Isaac, 24, 25, 26, 27

P

periodic table, 20, 21
phosphorus, 19
physicists, 24
protons, 16, 17, 19

S

Saturn, 28
states of matter, 4
surface tension, 9

V

volume, 8, 11

W

water, 4, 6, 7, 8, 9, 13, 14, 22
weight, 11, 29